629.4
BAK

Stepping Into Space

WEIGHTLESSNESS

by David Baker

THE AUTHOR — David Baker PhD, has been actively involved with NASA in the planning of the US space program. He has contributed to numerous aerospace and defense journals, and has written several books on space related subjects. Dr. Baker advises Independent Television News on American space operations and makes frequent TV appearances. He currently heads a consultant firm in space business development.

© 1986 Rourke Enterprises, Inc.

All rights reserved. No part of this book may be reproduced or utilized in any form or by any means, electronic or mechanical including photocopying, recording or by any information storage and retrieval system without permission in writing from the publisher.

Library of Congress Cataloging in Publication Data

Baker, David, 1944-
 Weightlessness.

 (Stepping into space)
 Summary: Discusses that condition which happens to astronauts and other things inside a spaceship in orbit.
 1. Weightlessness—Juvenile literature. [1. Weightlessness. 2. Astronautics] I. Title. II. Series:
Baker, David, 1944-   Stepping into space.
TL793.B235 1986   629.4'18   86-21934
ISBN 0-86592-970-X

PROPERTY OF
HOLY FAMILY SCHOOL
HILLCREST HEIGHTS, MD

Rourke Enterprises, Inc.
Vero Beach, FL 32964

06-320

On earth, we stand with our feet firmly planted on the ground. This is due to gravity. Gravity pulls us to the earth's center. A satellite circles the earth without drifting into space. Anything which circles a planet regularly is in an orbit. When astronauts fly into space, they become weightless. When they return, they feel the force of gravity. When a spaceship is in orbit, everything floats around it. There is no up or down as there is on earth. There are no floors or ceilings. This is confusing to humans, who are used to floors, walls, and ceilings.

In space there is no up or down and everything is weightless ▶

Engineers who design spaceships make use of every inch of inside space. They put lockers or containers of food or clothing anywhere. They know that an astronaut can simply float up or down to it. Of course, since everything in a spaceship is weightless, it is difficult to keep anything in one place. Objects float around. Loose items must be tied to a surface if they are to be kept in one place.

The Skylab Space Station was fitted out with many cubicles and lockers ▶

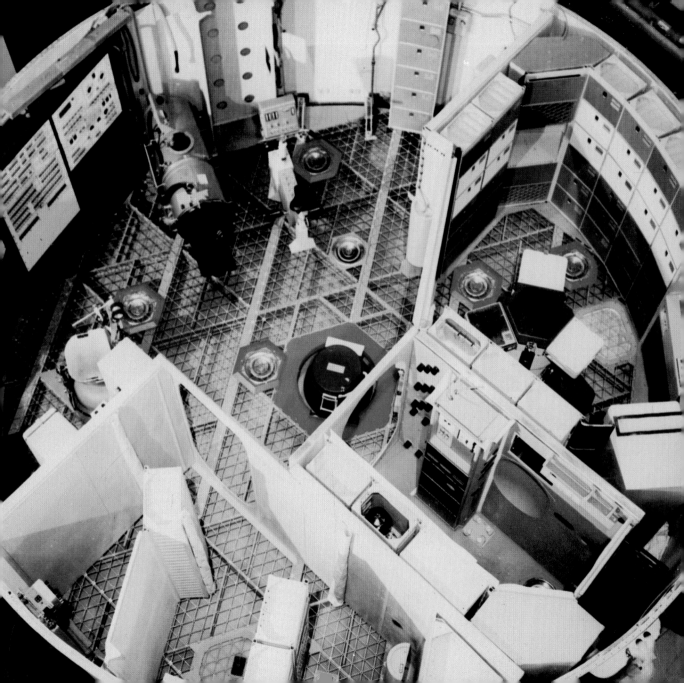

Although astronauts are weightless in orbit they must still eat, sleep and wash. Food would float away if it were left on plates. Special food containers keep it in one place. It is usually packed in meal trays on the ground. Most food items are in bite-size portions. Some food is sticky, so it can be held together. Crumbs could be dangerous in a spaceship. Tiny pieces of food could get into equipment or instruments and cause a problem. On board, food is kept in refrigerators.

Eating a meal in space needs special attention to prevent crumbs from floating ▶
around the cabin

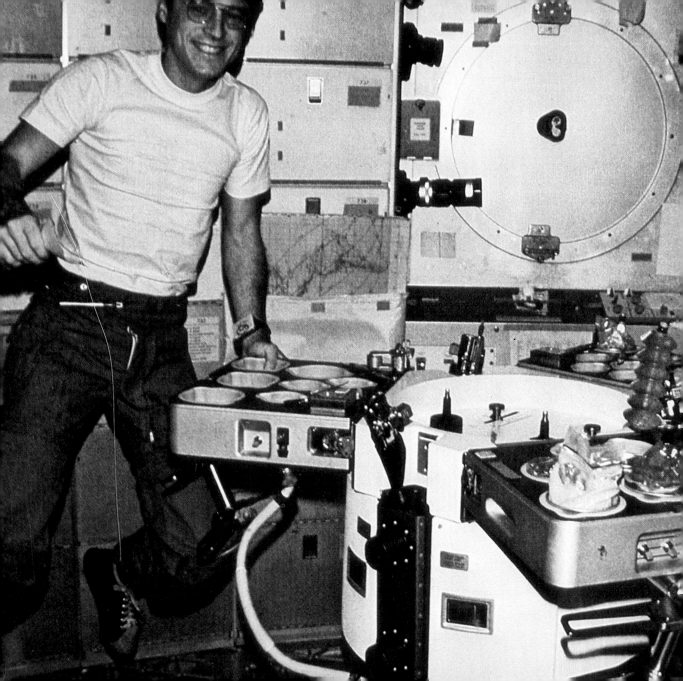

After a meal, cleaning up involves special equipment to keep the water from floating around. Some spaceships have showers with special curtains to hold the water. If left floating, water remains in bubbles. Like food crumbs, it could be dangerous if water got inside the equipment.

The washroom needs special design, with straps to keep the astronaut in ► *place and prevent him from floating away*

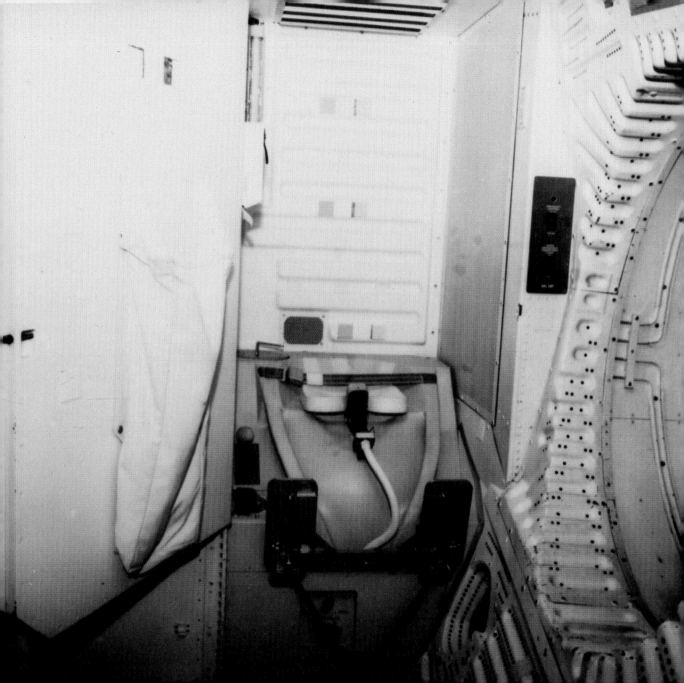

When astronauts go to sleep they have special bags to stop them from floating around and bumping into walls. Sleeping bags are attached firmly to the walls. Like campers in hammocks, astronauts have no need for beds.

Astronauts sometimes sleep in their spaceship like bats hanging from a wall ►

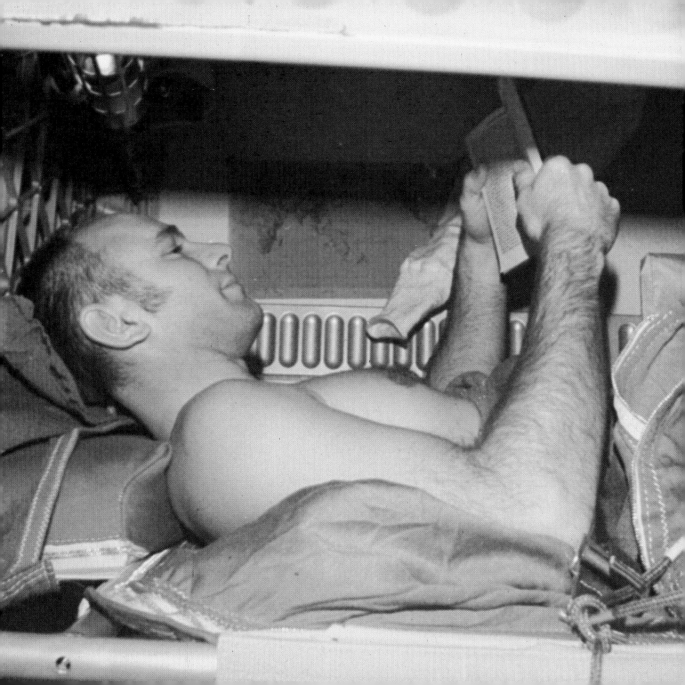

Living in space is different from living in a house on earth. In space, weightlessness allows the astronaut to use the entire inside surface to stow things. He or she can float to any corner of the room. On earth, we only use that part of the room that we can reach with our hands.

It's easy to touch the ceiling when you're weightless in space ▶

In space, astronauts stow their food and clothing in special lockers. The lockers are placed around important equipment. This makes better use of more space. Using tools in weightlessness can be difficult. If an astronaut used a screwdriver, he would turn his body around the tool unless he fixed himself to the floor. Engineers have designed special tools for astronauts to use in space.

Future space stations will be laid out with lockers and cabinets on the floor ▶ and on the ceiling

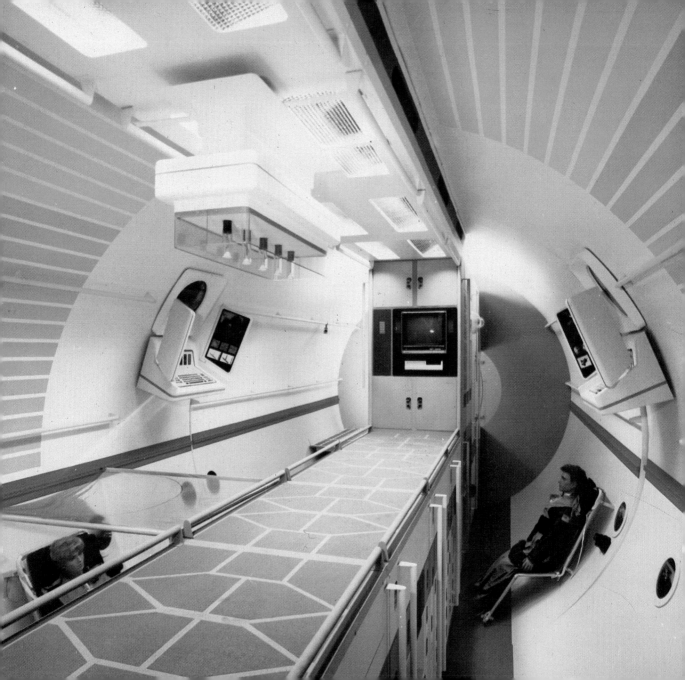

Moving around inside a spaceship is easy. Astronauts become used to gently pushing off a solid surface. They use their hands and feet to stop themselves. Their arms and legs become springboards.

Outside their spaceship, astronauts must wear special protective suits ▶

Outside the spaceship, there is nothing to stop an astronaut from floating off into space. Astronauts use special lifelines, called tethers, to attach themselves to the mother ship. They also use hand grips and special foot restraints to keep them in one position outside the spaceship. Shuttle astronauts can use special backpacks with jet thrusters to move around outside. If they are attached to the spaceship, they can use oxygen from the cabin fed along a lifeline. If they are not attached to the spaceship, they must carry the oxygen in a backpack. With all this equipment, they become a separate satellite of the earth, moving around it freely.

A special backpack allows astronauts to move around freely outside the shuttle ▶

Not everything about weightlessness is pleasant. Weightlessness gives the human body quite a shock. It is not used to a lack of gravity. The heart is a muscle and needs exercise to keep it fit. That means the astronaut must do daily exercise as he or she would on earth. The leg and arm muscles must also get exercise.

Astronauts must train to make themselves fit for space travel ▶

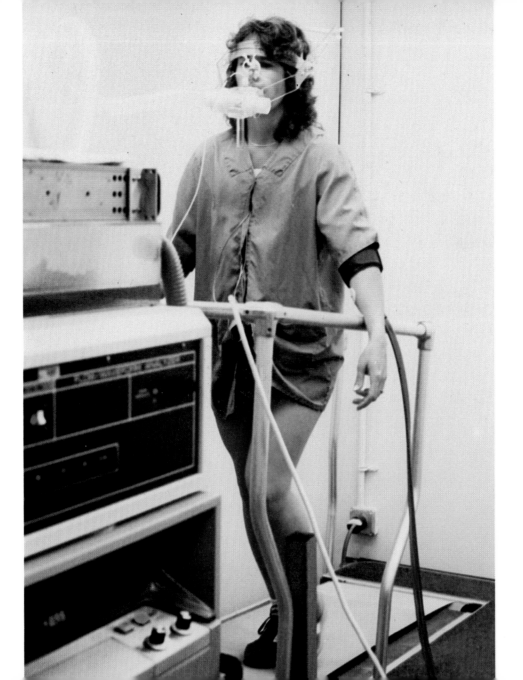

Special equipment is carried in the spaceship to keep the astronaut in good physical condition. Because he would float away, special restraint harnesses are necessary. The astronaut is held in position with freedom to exercise by these harnesses.

Medical tests carried out in space help scientists understand how the body ▶
reacts to weightlessness

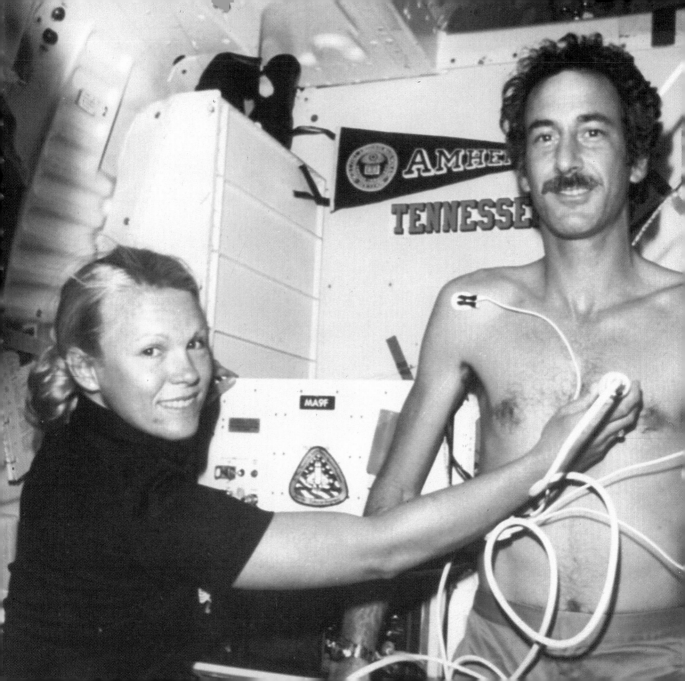

Sometimes the astronaut can use exercise equipment in a way impossible on earth. Using a bicycle exerciser to help arm muscles is one. On flights longer than a few weeks, astronauts' bones lose strength. Returning from weightless flights of a year or more, astronauts would need special bed rest. That would help put back the strength in their bones.

Daily exercise in space is necessary to keep fit for the stresses of earth gravity ▶
when they return

Weightlessness can allow people to do things impossible on earth. An astronaut can hold a 2,000 lb. satellite above his head with ease. He can lift heavy equipment inside the cabin.

An astronaut holds up a 2,000 lb. satellite with ease in the weightlessness of ▶ space

With the help of a mechanical arm he can move huge pieces of cargo. If used for scientific purposes, weightlessness can be useful. New medicines can be made in space which cannot be made on earth. In the absence of gravity, liquids are not pulled down to the bottom of a container. Scientists can prepare medicine in a way that is impossible within earth's gravitational pull.

A large robot arm helps an astronaut work outside his spaceship ▶

Some scientists are experimenting with plants. They believe astronauts on flights to other planets will have to provide their own food. Some plants grow better in weightlessness than they do on earth. One day, weightless grown fruit and vegetable gardens may support colonies living and working in space.

GLOSSARY

Astronaut   A man or woman who travels beyond the earth's orbit.

Gravity   The force that tends to move objects toward the center of a larger body like the earth or the sun.

Satellite   A natural or man made object in orbit around a major body like the earth or the moon.

Spaceship   A manned or unmanned vehicle designed for orbital flight, to be taken to the surface of another world or to fly freely in space.

Tether   A lifeline connecting the astronaut to the mother spaceship.

Weightlessness   The state or condition of being without weight due to the motion of a satellite or spaceship with respect to the body around which it is orbiting.

PROPERTY OF
HOLY FAMILY SCHOOL
HILLCREST HEIGHTS, MD.